Imitating *Nature*

From PINECONES to Cool **Clothes**

Other books in this series include:

From Barbs on a Weed to Velcro
From Bat Sonar to Canes for the Blind
From Bug Legs to Walking Robots
From Gecko Feet to Sticky Tape
From Insect Wings to Flying Robots
From Lizard Saliva to Diabetic Drugs
From Penguin Wings to Boat Flippers
From Spider Webs to Man-Made Silk

Imitating *Nature*

From PINECONES to Cool Clothes

Toney Allman

KIDHAVEN PRESS
An imprint of Thomson Gale, a part of The Thomson Corporation

THOMSON
GALE

Detroit • New York • San Francisco • San Diego • New Haven, Conn. • Waterville, Maine • London • Munich

For more information, contact
KidHaven Press
27500 Drake Rd.
Farmington Hills, MI 48331-3535
Or you can visit our Internet site at http://www.gale.com

LIBRARY OF CONGRESS CATALOGING-IN-PUBLICATION DATA

Allman, Toney.
   From pinecones to cool clothes / by Toney Allman.
      p. cm. — (Imitating nature)
   "Discusses how scientists have developed self-cooling clothing based on properties
      found in pinecones"—Provided by publisher.
   Includes bibliographical references and index.
   ISBN 0-7377-3490-6 (hard cover : alk. paper) 1. Textile fabrics—Technological
      innovations. I. Title. II. Series.
TS1767.A48 2006
677—dc22
                                                                  2005019840

Printed in the United States of America

Contents

CHAPTER 1

Surprising Pinecones

Almost everybody knows what a pinecone is, but many people do not know that pinecones open and close in response to the weather. Julian Vincent, a scientist at the University of Reading in England, was very interested in how pinecones worked. He wanted to invent cloth that could respond to the weather just like pinecones.

What Pinecones Do

Pinecones grow on pine trees, which are also called **conifers**. Conifer means "cone bearing," and every conifer has cones. Conifers live all around the world, but most of them grow in the world's **temperate** regions, such as North America, Europe, and Asia. Temperate environments are those that have both warm and cold seasons. Conifers grow best in regions that have four seasons.

Cones are the way conifers protect their seeds from harsh weather. Cones are seed holders for pine trees.

The hard, woody scales of the pinecone hold the seeds of life for the pine tree.

Scaly Seed Holder

1 Young pinecones stay tightly closed when the weather is cold and wet.

2 Pine seeds develop inside the young, closed cones.

3 When the weather is warm and dry, the cone's woody scales open up.

4 Each scale has two seeds. The cone releases the seeds when it is fully open.

The cones are made up of many hard, woody **scales** that hold pine seeds inside. At the base of each scale are two seeds that will be released from the cone when it opens up. Cones come in different sizes, depending on the kind of conifer on which they grow. Pinecones can vary in size from ¼ inch (0.64cm) long to 2 feet (61cm) long.

No matter how big or small, pinecones stay tightly closed as long as the weather is wet and cold. Pine seeds would die if they were exposed to wintry conditions. In the spring, pinecones remain tightly closed because they absorb water from the tree while they are growing. When summer days arrive or the ripe cones fall off the tree, the scales of the cones dry out and bend open. The seeds pop out from the bases of the scales, and thousands of pine seeds scatter everywhere.

Tough and Amazing

Conifers can survive droughts, freezing winters, high winds, and fire. Some conifers are record breakers, too. The largest living thing on Earth is a conifer called the sequoia (right). It can be almost 300 feet (91.4m) tall, with a trunk 35 feet (10.7m) thick. Another conifer, the bristlecone pine, is the oldest living thing on the planet. Bristlecone trees can live as long as 5,000 years.

Some pinecones, such as the Monterey cones of California, open only when it is very hot or there is a forest fire. The heat of a forest fire dries out the cone and forces it open. Fire also clears the forest floor, so that the pine seeds drop onto clear soil and have a good chance to grow. Without a fire, Monterey pinecones will pop open only when the temperature is about 90°F (32.2°C). This is nature's way of spreading seeds to many different places and ensuring that some survive to become new Monterey pine trees.

Gathering Cones for Science

Monterey pines have large cones, about 3–5 inches (7.6–12.7cm) long. This makes them easy for scientists to study. In 1995 Vincent and his students collected pinecones from the big Monterey pine tree that grew next to their university's library. Monterey pines are not native to England, but long ago, in 1833, some were brought to England

The Monterey pine grows on the California coast. Its cones (inset) open only in very high heat.

and planted there. Vincent and his team took their cones into the laboratory to study how the scales were able to open up and release their seeds in hot weather.

Stretching and Shrinking

Vincent tore apart the pinecones and examined their scales. He could see that each scale is made up of many woody **fibers**, arranged in two different layers. The fibers of the outside layer run in a different direction from the inside layer. When the pinecone fibers are dry, the scales are naturally open. When the scales get wet, however, the fibers of the outer layer swell, or **expand**, more than the inner fibers do. This forces the scales to bend inward so the pinecone closes up. It stays tightly closed until it dries out again.

Vincent thought the cones were very tricky. He wondered if he could imitate nature's fiber arrangement. Perhaps he could make cloth that opened and closed in response to temperature and dryness. Clothes made with such cloth could keep people comfortable in any kind of weather.

An image created with a special microscope shows the swirled inner fibers within a section of pinecone.

Surprising Pinecones

A special microscope shows thin, stringy threads of wool and an even more detailed view of angora wool fibers (magnified 1,000 times in inset).

CHAPTER 2

From Scales to Cloth

In 1995 the British government gave Vincent and his scientific team money for their study of pinecones. For three years, the scientists experimented with how to make fibers for cloth that could open and close like pinecone scales.

Fibers for All Climates

All cloth is made of fibers. Fibers are thin, stringy threads of material. Fibers of different materials can be woven together to make cloth. The fibers in pinecone scales are not long enough to make cloth, but fibers from some plants, such as cotton, make very good cloth. Other clothing fibers come from animals, like wool from sheep. Scientists have also made artificial fibers, such as nylon, out of wood, plastic, or even metal. Vincent had to experiment with many different kinds of fibers before he found some that could open and close the way pinecone fibers do.

The cross-shaped fibers that make up nylon are visible under a microscope.

Vincent's plan was to make cloth that could change shape, depending on whether its fibers were wet or dry. It would shrink when dry and expand a little when it was wet. Only the changed texture would be visible, but microscopic openings in the cloth would cool off a wearer.

Vincent knew that people sweat when they are hot. When they cool off, in a breeze, for example,

Young cones remain tightly closed in cool, damp weather (below) while hot, dry weather has opened the cone at left.

they stop sweating, and their skin dries. Clothing that expanded when people perspired would let them cool off. Then when their skin was dry, the clothing would shrink again. Vincent's cloth would work the opposite way from pinecones. Pinecones open when dry and close when wet. Vincent's fibers would open when wet and close when dry. However, the idea was the same: the fibers in the cloth would imitate the way pinecone fibers work.

Wearable Flaps

Vincent and his team got an idea. They would cut microscopic flaps in specially made cloth. The fibers of the cloth would have to be able to absorb moisture, expand when they were wet, and push the flaps outward so they would open. Then, when the fibers dried out, they would shrink, or **contract**, so that the flaps lay flat again. The flaps would not work until they found the right kind of fiber.

Acting Intelligently

Things that can change their shape or length because of a change in their environment are called "smart materials." Pinecones are smart materials because they open and close in response to moisture. Some ferns and mosses are smart materials, too, using moisture in their environment to time their seed release. Julian Vincent and Veronika Kapsali are studying these plants, also, to help them design cool clothing.

A fern uncoils in response to weather.

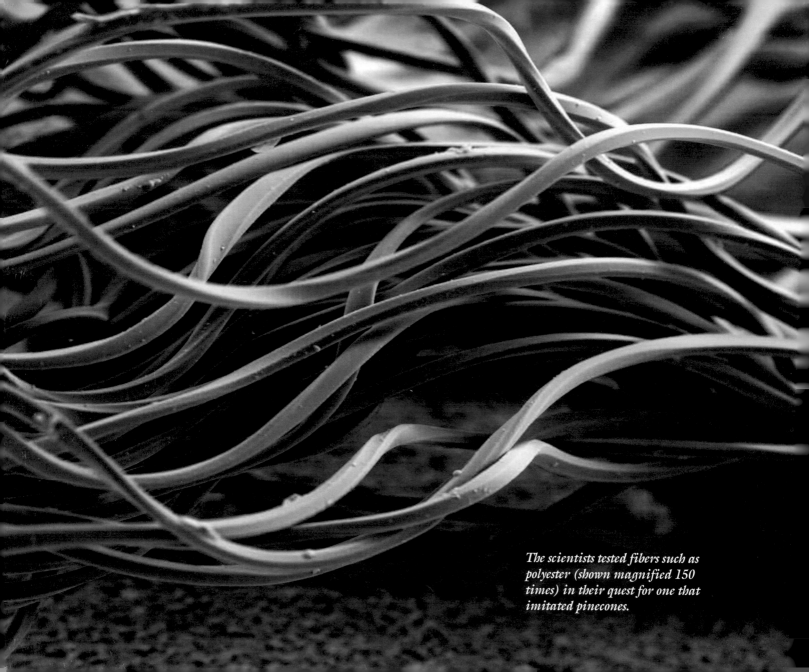

The scientists tested fibers such as polyester (shown magnified 150 times) in their quest for one that imitated pinecones.

The scientists experimented with many different kinds of fibers. From wool to cotton to man-made fibers, they searched for a fiber that behaved like pinecone fibers. Finally, they found a fiber that responded to moisture the way they wanted. They are keeping this fiber a secret for now. They wove small pieces of cloth with the fiber, and they cut tiny flaps into it under a microscope. Each flap was only ⅕,₀₀₀ of an inch (0.0005cm) wide, so tiny that it was invisible.

Vincent breathed on one small piece of the cloth. The warm moisture of his breath was like perspiration on skin. The flaps of the cloth opened, letting a cool breeze pass through the cloth. When the fresh air dried the fibers, the flaps closed up again.

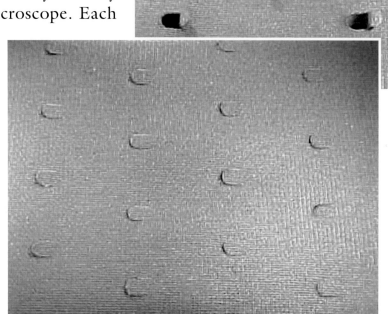

Tiny flaps open (above) and close (left) in Julian Vincent's cloth.

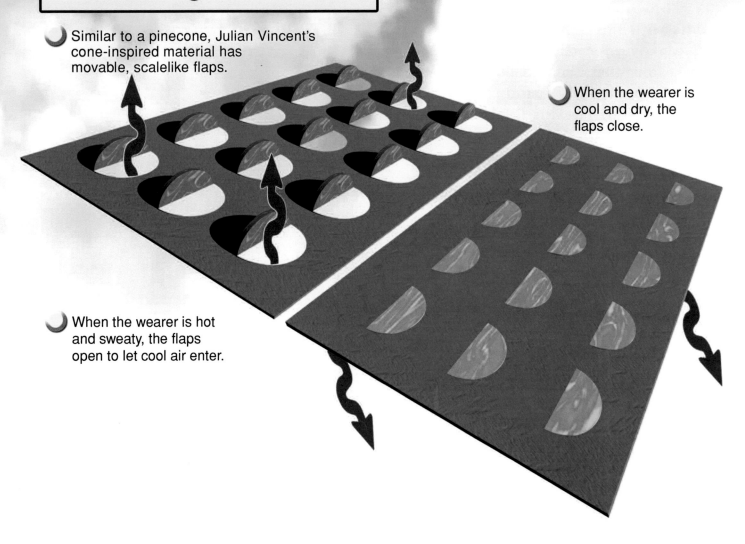

Amazing Material

Similar to a pinecone, Julian Vincent's cone-inspired material has movable, scalelike flaps.

When the wearer is hot and sweaty, the flaps open to let cool air enter.

When the wearer is cool and dry, the flaps close.

Vincent's idea had worked. Clothes made with these expanding and contracting fibers would shrink to keep the wearer warm when the weather was cool. When the weather was hot, the flaps would open and cool off the wearer.

No Good After All

Vincent's cloth was a success, but then came a terrible problem. The British government was paying for Vincent's research because they wanted cool clothing for the military. However, soldiers' uniforms have several layers. Vincent's cloth would not work under layers of clothes because the little flaps stick out a bit from the cloth. The flaps could not push open under another layer of clothing. They would be trapped by the top layer.

The military thought the cloth was a failure, and the government would not pay for Vincent's research anymore. For a time, it seemed that Vincent's work had come to an end.

The Beginning of a Journey

One winter, Veronika Kapsali did an experiment as she rode and walked throughout London. Again and again, she measured the temperature of her skin and the air temperature with special instruments. She wrote down how she felt as she moved around. When she studied her results, she discovered that it is difficult to dress comfortably for a whole day because her activities changed during different times of the day. If she had to run to catch a bus, she got too hot. Other times, when she sat still, she was chilly. Her experiment inspired her to search for a way to make cool clothing a reality.

CHAPTER 3

Pinecone Clothes for the Future

In 2004 Vincent's pinecone cloth was chosen for display at Expo 2005 in Aichi, Japan. The Expo is a worldwide exhibit about learning from nature and protecting the environment. When people, businesses, and governments heard about Vincent's pinecone clothes they became excited. Vincent and his team were able to seek more research money to start experimenting again. When they succeed in making these clothes, people will have smart, cool clothing for everyday use.

Starting Over

Vincent moved to a new laboratory at the University of Bath. There, he was joined by Veronika Kapsali, a fashion designer and **textile** expert at the London College of Fashion. The two researchers have to come up with a completely new fiber that will allow them to develop a pinecone cloth that does not rely on flaps for cooling and warming.

Visitors to Expo 2005 in Japan got a chance to learn about Julian Vincent's pinecone cloth for the first time.

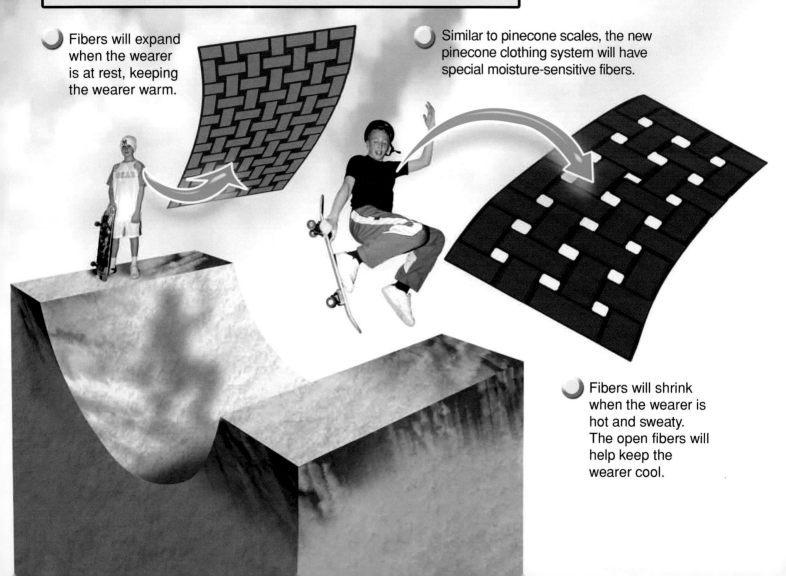

Improved Pinecone Clothes

Fibers will expand when the wearer is at rest, keeping the wearer warm.

Similar to pinecone scales, the new pinecone clothing system will have special moisture-sensitive fibers.

Fibers will shrink when the wearer is hot and sweaty. The open fibers will help keep the wearer cool.

Vincent got an idea for a new pinecone cooling system. This time, instead of flaps that open and close, his cloth fibers will expand to keep air out and shrink to let air in. Vincent says these conelike fibers would be open, just as pinecones open when they are dry. Then, as the fibers absorb moisture or sweat, the fibers would shrink, and the space between them would get larger throughout the material. These microscopic "holes" in the clothing would let air in and cool off the wearer. When the fibers dried out, they would expand again, closing the holes, so the wearer would be warm.

Making the Fiber Work

Underneath the pinecone fibers would be another layer. Kapsali says this layer would be very thin and transparent. It would protect the wearer from rain that might get through the holes in the top layer. Clothes made from cloth like this would cool and protect at the same time.

All Kinds of Smart Clothing

Pinecone clothing will help people stay cool, but a musical jacket from Levi Strauss & Co. will help people have fun. The jacket is embroidered with special threads that the wearer can touch to play music. Scientists at Georgia Tech have invented a shirt with computer technology sewn into it. This shirt can check the wearer's heartbeat and breathing. Someday, it may be used to monitor soldiers on the battlefield or to help doctors know when sick people need medical attention.

Special threads make this jacket a musical masterpiece.

Vincent and Kapsali have not found a fiber that works the way they want it to, so they are inventing their own. Vincent says it may be another year before that happens. At that time, Kapsali will weave and design clothing from it that will be cool, practical, and fun to wear.

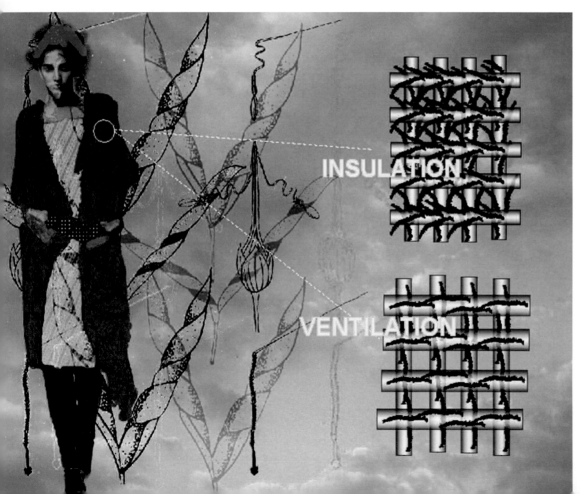

A drawing by Veronika Kapsali shows the tiny holes that would provide both warmth and cooling for the wearer.

INSULATION

VENTILATION

Cool Clothes for Everyone

When the cloth is perfected, the military would have cool clothing that works even under layers of clothes. Soldiers in hot places, such as deserts, would be able to cool off without removing any of their gear. When night falls, and the temperature is cool, the clothing would close up and keep the soldiers warm.

Athletes might want cool clothing, too. People who play sports, run and jog, or bicycle long distances need clothing that responds when they perspire. Instead of removing or changing clothes when they are too hot, these athletes can wear smart clothing that lets them cool off automatically.

Clothing that warms and cools the body might one day benefit runners, cyclists, and other athletes.

Other people could benefit from cool clothing, as well. Firefighters have to wear heavy, protective equipment when they fight fires. The closer they get to the fire, the hotter they get. Firefighting clothes made from smart fibers would be a great advantage. Elderly people, who may have trouble cooling off on summer days, can get sick from the heat. Someday, they could wear pinecone clothing both for comfort and safety. Young people may just enjoy the fun and convenience of pinecone clothes.

Thanks to the Pinecones

Cool clothing is a few years in the future, but when it arrives, everyone could have hats, pants, shirts, and coats that are both comfortable and nice to look at. When cool clothes become a reality, everyone's clothing will open and shut to adjust to the temperature, just as pinecones do.

Pinecone clothing could help firefighters on the job, keeping them safe and cool.

From Pinecones to Cool Clothes

Glossary

conifers: Cone-bearing trees, such as pines, spruces, and firs.

contract: Draw together or shrink.

expand: Increase in size or spread out.

fibers: Thin, threadlike structures that make up many natural materials.

scales: The hard, woody leaves, or shingles, which make up a pinecone. Pine seeds are held at the base of each scale.

temperate: Having temperatures that are neither extremely hot nor freezing cold throughout the year. The temperate zone of Earth has changing seasons.

textile: Cloth, or the materials that can be woven into cloth.

For Further Exploration

Books

Jason Cooper, *Pine Tree (Life Cycles)*. Vero Beach, FL: Rourke, 2003. Follow the pine tree as it grows to maturity, makes seeds, and reproduces in its natural environment.

Stephanie Maze, *I Want to Be a Fashion Designer (I Want to Be)*. San Diego: Harcourt, 2000. This book describes careers in fashion design with photos and stories about famous and not-so-famous clothing designers.

Web Sites

Coulter Pine Seed-Bearing Cone, Wayne's Word (http://waynesword.palomar.edu). See a photo of one of the world's largest pinecones.

Expo 2005 Aichi Japan (www-1.expo2005.or.jp/en). This is the official Web site for the world exposition in Japan. Explore the links to see the pavilions and exhibits, and learn about the mission of the Expo.

Nearctica: Native Conifers of North America (www.nearc
tica.com/trees/conifer/index.htm). Click the different links
to see many photos of different and unusual conifers and
their cones.

**Pine Cone Counting, British Columbia's Coastal Envi-
ronment** (www.educ.uvic.ca/faculty/mroth). Learn about
a mathematical relationship that pinecones share with other
plants by counting pinecone scales.

Textiles—Making Textiles, About.com (http://inventors.
about.com/library/inventors/bl_making_textiles.htm).
Learn about the history of making cloth from natural fibers,
as well as how clothing is made today. Follow the many links
to read about other textile topics, such as manufactured fibers
and how specific clothing items were invented.

Index

Picture Credits

Cover: (From left) Photos.com; Photos.com; Courtesy of Julian Vincent; Courtesy of Veronika Kapsali

About the Author

Toney Allman has degrees from Ohio State University and the University of Hawaii. She currently lives in Virginia by the Chesapeake Bay. Recently she has been tearing apart pinecones and examining their scales.